ISBN 978-1-332-28434-4
PIBN 10308942

1 MONTH OF
FREE
READING

at

www.ForgottenBooks.com

By purchasing this book you are eligible for one month membership to ForgottenBooks.com, giving you unlimited access to our entire collection of over 700,000 titles via our web site and mobile apps.

To claim your free month visit: www.forgottenbooks.com/free308942

English
Français
Deutsche
Italiano
Español
Português

www.forgottenbooks.com

Mythology Photography **Fiction**
Fishing Christianity **Art** Cooking
Essays Buddhism Freemasonry
Medicine **Biology** Music **Ancient
Egypt** Evolution Carpentry Physics
Dance Geology **Mathematics** Fitness
Shakespeare **Folklore** Yoga Marketing
Confidence Immortality Biographies
Poetry **Psychology** Witchcraft
Electronics Chemistry History **Law**
Accounting **Philosophy** Anthropology
Alchemy Drama Quantum Mechanics
Atheism Sexual Health **Ancient History**
Entrepreneurship Languages Sport
Paleontology Needlework Islam
Metaphysics Investment Archaeology
Parenting Statistics Criminology
Motivational

TEST-RETEST CORRELATIONS AND
THE RELIABILITY OF COPY TESTING

Alvin J. Silk*

WP 917-77   March 1977

TEST-RETEST CORRELATIONS AND
THE RELIABILITY OF COPY TESTING

Alvin J. Silk*

WP 917-77   March 1977

* This paper was written while the author was Visiting Professor, European
  Institute for Advanced Studies in Management, Brussels. Thanks are due
  to Alain Bultez and Christian Derbaix for valuable discussions of this
  topic and to Donald Morrison for helpful comments on an earlier draft.

INTRODUCTION

Reliability is one of the basic criteria about which advertisers seek information in evaluating copy testing methods [7, p. 15]. Gross [15, 16] has shown how the reliability of the pre-testing procedure employed affects a firm's policy with respect to the creation and testing of copy alternatives. The measure of reliability most often reported in the advertising literature is a test-retest correlation coefficient [1, 4, 12, 28, 37]. However, practically no attention has been paid to the requirements which test-retest data must satisfy in order for a correlation derived from them to represent an interpretable indicator of reliability. This paper is concerned with the commonplace practice of using ad hoc collections of hetereogeneous test-retest data as a basis for computing correlations which are then presented as measures of the reliability of copy testing methods.

The presentation proceeds as follows. First, the assumptions under which a test-retest correlation corresponds to the concept of reliability used in psychological measurement are examined. Next, the consequences of certain types of departures from these assumptions are investigated and some checks are suggested to assess the consistency of test-retest data with conditions necessary for a test-retest correlation to be taken as a theoretically meaningful index of reliability. Applications of the consistency checks are illustrated with recall scores from copy tests of television commercials. Finally, doubts and limitations surrounding many available reliability estimates are discussed.

THE CONCEPT OF RELIABILITY

We begin by considering the concept of reliability found in the
psychometric literature concerned with "parallel" or "equivalent" measures.
[17, chap. 14; 24, chapts. 2, 3 and 7]. The underlying measurement
model is formally equivalent to that employed by Gross [16] with reference to copy testing and to that presented by Morrison [27] with
reference to marketing research test instruments in general.

Assume:

$$(1) \quad O_{it} = \theta_i + \epsilon_{it} \quad ,$$

where : O_{it} = the observed copy test score for the ith advertising
alternative on the tth replication of the test,
$i = 1,\ldots,n$ and $t = 1,\ldots,K$,

θ_i = the "true" copy test score for the ith advertising
alternative,

ϵ_{it} = error component in the observed copy test score
for the ith advertising alternative arising on the
t^{th} replication of the test.

Assume further that θ_i and ϵ_i are random variables, independent
normally distributed according to:

$$(2) \quad f(\theta_i) \sim N(\mu_\theta, \sigma_\theta^2) \quad ,$$
$$(3) \quad f(\epsilon_i) \sim N(0, \sigma_\epsilon^2) \quad ,$$
$$(4) \quad \rho_{\theta\epsilon} = 0$$

where:

μ_θ =   mean of the distribution of the true copy test scores,

σ_θ^2 =   variance of the true copy test scores,

σ_ϵ^2 =   variance of the errors of observation,

$\rho_{\theta\epsilon}$ =   correlation between the true copy test scores and the
errors of observation.

As is well-known, it follows from the above assumptions that the expected value of the observed scores is equal to the true score mean:

(5) $E(O_i) = \mu_\theta$

and the variance of the observed scores, σ_o^2 is:

(6) $\sigma_o^2 = \sigma_\theta^2 + \sigma_\varepsilon^2$.

The reliability coefficient,R, generally used in the psychometric literature [e.g., 24, p. 61] is defined as the ratio of true score variance to observed score variance:$\underline{\frac{1}{}}$

(7) $R = \sigma_\theta^2 \,/\, \sigma_o^2$

       $= 1 - \sigma_\varepsilon^2 \,/\, \sigma_o^2$.

Another interpretation of reliability frequently mentioned in psychometric discussions is that R equals the square of the correlation between observed and true scores. The derivation of this relationship may be found in [24, p. 57].

TEST-RETEST CORRELATIONS

The great bulk of the available empirical evidence bearing on the reliability of copy testing procedures has been developed from studies which have concerned themselves with the "stability" or "reproducibility" of scores without bothering to specify a formal model of the underlying process whereby these measurements are generated. What is most often discussed is the "consistency" of the scores obtained when the same set of ads are tested with the same method on two separate occasions -- typically with two separate samples of respondents drawn at different

points in time. To summarize the degree of stability or consistency
exhibited by the two sets of observed scores, the value of the ordinary
product-moment correlation coefficient is calculated for such data.
Let us now consider how such correlations may be related to the concept of reliability outlined above.

Suppose we test and retest n advertisements (each with two independent
but equivalent groups of respondents) and thereby obtain two sets of
copy test scores generated by the basic measurement model defined by (1) -
(4) above. Referring to Equation (1), let t = 1 to denote the initial
"test" and t = 2 to denote the "retest", then:

(8) $O_{i1} = \theta_i + \epsilon_{i1}$

  $O_{i2} = \theta_i + \epsilon_{i2}$

  $i = 1,\ldots,n$

Let us also assume:

(9) $\sigma^2_{\epsilon_1} = \sigma^2_{\epsilon_2} = \sigma^2_{\epsilon}$

and

(10) $\rho_{\epsilon_1 \epsilon_2} = 0$

Now consider the correlation coefficient between the two sets of observed
scores, O_{i1} and O_{i2} :

(11) $\rho_{O_1 O_2} = \dfrac{\text{Cov}(O_{i1}, O_{i2})}{[\sigma^2_{O_1} \sigma^2_{O_2}]^{1/2}}$

where:

$\text{Cov}(O_{i1}, O_{i2})$ = Covariance of the observed test and retest scores,

$\sigma^2_{o_1}$, $\sigma^2_{o_2}$ = Variances of the observed test and retest scores, respectively .

The covariance of the observed scores is:

(12) $\text{Cov}(O_{i1}, O_{i2}) = E[O_{i1}O_{i2}] - E(O_{i1})E[O_{i2}]$.

However, since from (2) and (8) it follows that:

(13) $E(O_{i1}) = E(O_{i2}) = \mu_\theta$,

and since we assume $\rho_{\theta\varepsilon_1} = \rho_{\theta\varepsilon_2} = 0$, $E(\varepsilon_{i1}) = E(\varepsilon_{i2}) = 0$, and $\rho_{\varepsilon_1\varepsilon_2} = 0$, it is readily shown that (12) reduces to:

(14) $\text{Cov}(O_{i1}, O_{i2}) = \sigma^2_\theta$.

This result is a key element in the psychometric theory of parallel measurements: "The true score variance (an unobservable quantity) is equal to the covariance between parallel measurements (a potentially observable quantity)" [23, p. 59].

From (6) and (9) it also follows that:

(15) $\sigma^2_{o_{,1}} = \sigma^2_{\rho_2} = \sigma^2_\theta + \sigma^2_\varepsilon = \sigma^2_o$.

Substituting (14) and (15) for the numerator and denominator, respectively, of (11), we find:

(16) $\rho_{o_1o_2} = \sigma^2_\theta / \sigma^2_o = R$.

Thus, under the assumptions stated, the correlation coefficient between the observed test and retest scores is equal to the ratio of the true and observed score variances and therefore is equivalent to the usual reliability coefficient defined above in (7)$\underline{2/}$.

The Importance of Equivalent Test-Retest Conditions

How do departures from the assumptions underlying the above measurement model affect the interpretability of a test-retest correlation as a reliability coefficient? Attention here will be focused upon one likely source of problems, failure to satisfy certain stability conditions in the distributions of scores across replications. Note from (8) that the true scores of the individual ads or commercials are assumed to be identical in the test and retest situations while their error components may vary, subject the restriction that for each replication the distribution of the errors across ads or commercials has the same (zero) mean (3) and variance (9).

It is important to inquire about the consequences of violating these assumptions because it seems to have generally been taken for granted that a correlation computed from virtually any set of test-retest data can be interpreted as a reliability coefficient in a straightforward manner. As others have noted,[33, p.56], only rarely have studies been undertaken for the expressed purpose of assessing the reliability of copy testing methodologies and where controls have been exercised to achieve comparability in test and retest conditions. The great bulk of the test-retest correlations now available for copy testing methods have been developed from ad hoc bodies of data assembled from previous studies where commercials had been retested for any of a number of reasons. The requirements of such work are not very likely to permit, much less demand, equivalent test-retest conditions. A case in point are "wear out" studies which are a major source of test-retest data for copy testing systems. A new commercial is typically first tested with

a sample of respondents who have viewed it once or perhaps a few times. Following this it continues to be aired regularly for some period of time and then a retest is carried out to check on the durability of its performance. Consideration of a hypothetical but concrete example will serve to illustrate the difficulties that may be encountered in using such data to assess reliability.

Suppose that the initial test measurements for a set of commercials are taken after each one has been broadcast only once and then the second or retest measures are obtained after the commercials have each been aired X additional times. Now assume that as a result of this repeated exposure, a systematic shift occurs in the true values of the response measured. What effect does this true score shift have on the test-retest correlation as a reliability coefficient? The answer depends upon the nature of the change involved. This may be demonstrated by examining the consequences of two different types of shifts in true scores that might occur: (a) a uniform additive change, and (b) a uniform proportional change.

A. Uniform Additive Change. Let us retain all the preceding assumptions (1-4) and (9-10) but represent this type of change in the retest measures by replacing (8) with:

(8A) $O_{i1} = \theta_i + \varepsilon_{i1}$

$O'_{i2} = \delta + \theta_i + \varepsilon_{i2}$,

where:
δ = additive shift parameter.

That is, we assume that the impact of the repeated exposure is to increase (or decrease) the retest score for each commercial from its initial test level by some fixed amount, δ . Comparing the distributions of the test and retest scores, we can easily show that their means would no longer be the same:

(13A) $E(O_{i1}) = \mu_\theta \neq E(O'_{i2}) = \delta + \mu_\theta,$

but the covariance of the observed scores would still be equal the
variance of the true scores.

(14A) $Cov(O_1, O'_2) = \sigma_\theta^2,$

and the equality of the variances of the observed test and retest scores
would be preserved:

(15A) $\sigma_{O_1}^2 = \sigma_\theta^2 + \sigma_\epsilon^2 = \sigma_{O'_2}^2$

Hence it can be seen that the additive shift constant does not enter
into either the numerator (14A) or the denominator of the test-retest
correlation (11) and so it would still represent a reliability coefficient:

(16A) $\rho_{O_1 O'_2} = \dfrac{\sigma_\theta^2}{\sigma_\theta^2 + \sigma_\epsilon^2} = R.$

B.   Uniform Proportional Change. Assume that the retest scores
shift. from their initial test levels so that the rate of change
is the same for all the commercials. That is, instead of (8), assume:

(8B) $O_{i1} = \theta_i + \epsilon_{i1}$,

$O''_{i2} = \alpha\theta_i + \epsilon_{i2}$,

where:

        α = proportional shift parameter .

Such a change process is somewhat more realistic than a uniform additive
shift in that it is consistent with the use of exponential response
functions to describe the effects of repeated exposure or over-time
decay or forgetting. Again comparing the test and retest scores, we
find that their means are unequal:

(13B) $E(O_{i1}) = \mu_\theta \neq E(O''_{i2}) = \alpha\mu_\theta$,

and the covariance of the observed scores is now given by:

(14B) $Cov(Oi_1, O''_{i2}) = \alpha\sigma^2_\theta$

The variances of the observed test and retest scores also cease to be equal:

(15B) $\sigma^2_{O_1} = \sigma^2_\theta + \sigma^2_\epsilon \neq \sigma^2_{O''_2} = \alpha^2\sigma^2_\theta + \sigma^2_\epsilon$

and therefore the test-retest correlation becomes:

(16B) $\rho_{O_1 O''_2} = \dfrac{\alpha\,\sigma^2_\theta}{[(\sigma^2_\theta + \sigma^2_\epsilon)\,(\alpha^2\sigma^2_\theta + \sigma^2_\epsilon)]^{1/2}}$.

Thus it can be seen that the test-retest correlation is affected by the proportional shift parameter, α and no longer directly represents a ratio of true to observed score variance. In this case, the test and retest scores would, in fact, have different reliabilities even though their error variances were identical:

$$R_1 = \frac{\sigma^2_\theta}{\sigma^2_\theta + \sigma^2_\epsilon}$$

$$R_2 = \frac{\alpha^2\sigma^2_\theta}{\alpha^2\sigma^2_\theta + \sigma^2_\epsilon}$$

Here the test-retest correlation (16B) would represent the geometric mean of these two separate reliability coefficients:

$$\rho_{O_1 O''_2} = [R_1 R_2]^{1/2} .$$

Also, the _square_ of the test-retest correlation provides a lower bound for the two unobserved reliabilities because the values of all three of these quantities must lie between 0 and 1:

$$R_1 > \rho^2_{o_1 o''_2} < R_2 \quad .$$

The above examples illustrate how the meaning of a test-retest correlation can be obscured when a certain kind of departure from the assumed stability conditions occurs. The preceding analysis deals with situations where the true scores of all the ads or commercials studied underwent the same type of systematic change. Of course, a misleading picture of reliability could also be obtained from a test-retest correlation where each ad had been affected by a different or idiosyncratic shift. Suppose a set of ads are tested and retested by a copy testing method that produced scores entirely free of random measurement error. However, between the two tests, assume the scores for the ads change but in a non-uniform manner -- i.e., some increase and others decrease. It follows that the test-retest correlation for such data would be less than unity and therefore one might be led to conclude that the measures were fallible when, in fact they were perfect.

Diagnostic Test

Application of an appropriate significance test would, in principal, detect the problematical types of shifts in the distributions of scores discussed above and thereby alert the analyst to a violation of one or another of the assumptions necessary for a test-retest correlation to be taken as a measure of reliability. The equality of test and retest variances (15) implied by the measurement model provides a basis for a useful consistency check. That is, if a particular set of test-retest data were generated in conformity with the assumptions (1-4) and (8-10) then subject to sampling fluctuations, the variances of the observed test and retest scores should be found equal. With normally distributed measures, the variance of observed scores for a random sample of ads/commercials copy tested by some method will be an unbiased estimate of the corresponding population parameter and hence the null hypothesis of equal test and retest variances ($\hat{\sigma}^2_{o_1}$ and $\hat{\sigma}^2_{o_2}$, respectively) can be tested:

$$(17) \quad \hat{\sigma}^2_{o_1} = \hat{\sigma}^2_{o_2}$$

Rejction of this hypothesis would provide evidence to contradict the assumptions under which a test-retest correlation can serve as a reliability coefficient in the sense of (7). Unequal observed score variances implies that either the true or the error variances (or both) were not constant between the test and retest situation. The equality of observed score variances is, of course, a necessary but not sufficient condition for the assumed measurement model to hold.

Referring to the two types of uniform shifts examined in the preceding section, notice that the equality of test-retest variances holds in the case of an additive change (15A) where a test-retest correlation does represent a reliability coefficient but this condition is violated when a proportional change occurs (15B) and a test-retest correlation does not have a simple reliability interpretation. As for uniform proportional shifts, unequal test-retest variances could also arise when idiosyncratic, non-compensating changes take place in the true scores of the individual ads/commercials.

If the observed test and retest variances are found to be equal, the next step is to investigate the equality of the test and retest means. The latter condition is expected when the basic parallel measurement hold holds. If the test and retest variances are equal but their means are unequal, then the question arises as to whether or not it is meaningful to assume the situation is equivalent to Case B above where a uniform additive shift results in a difference between the test and retest means but a test-retest correlation still represents a reliability coefficient. Examples of how these checks and some related types of analysis may be utilized are given in the next section.

ILLUSTRATIVE EXAMPLES

Below we examine two small bodies of data similar in kind to that which test-retest correlations reported in the advertising research literature have often been based. Both examples involve recall scores from "on-air" tests of television commercials. The analyses presented are intended to illustrate the use of some statistical methods but given the limited scope of the data bases examined, no broader interpretation of the results obtained is warranted.

Case I

The first example is based upon a small set of data found in an unpublished paper by Heller [18] and consists of "related recall" scores obtained from separate test-retest studies of four different commercials for the same unidentified brand. Heller mentions that the four test-retest studies were all done three days apart but other details about the sampling and test conditions were not reported.

Recall scores are proportions and as is well-known, the sampling variance of binomial proportion($\pi(1-\pi)/N$ where π is the true mean and N is the size of the sample of respondents) is dependent upon the mean. Therefore, in order to maintain consistency with the assumption of homoskedasticity (4), it becomes necessary to apply a variance stabilizing transformation to the original observations. The angular or arcsin transformation has this property [6, Chapt. 7] and is employed here. Since copy tests of different alternatives for the same brand are normally executed with roughly equal-sized samples, it would seem reasonable to assume that differences in the latter quantity are not likely to be an important source of heteroskedasticity here. Table 1 presents some summary statistics for both the original and arcsin transformed (degrees) observations -- i.e., the transformed score $Y = \sin^{-1}(X)^{1/2}$ where X is the original recall proportion.[3]

INSERT TABLE I HERE

This case is a rather striking example of how uncritical reliance on a test-retest correlation coefficient would lead to a misleading impression of reliability of the measures.[4] From Table 1 we see that the value of the test-retest correlation is .63 for the original observations and increase slightly to .71 after the arcsin transformation is applied. Taken at fact value this correlation would indicate a level of reliability that is about average when it is compared to other available on-air recall reliability

estimates [see, e.g., 29] although these coefficients are based on only four pairs of test-retest scores and neither is statistically significant (.40 > p > .20, 2 tail test). However, an examination of the other descriptive statistics in Table 1 reveals that the distributions of the test and retest score are quite dissimiliar. The retest mean score is almost 5 points below the initial level and more critically, the variance of the initial test recall scores is approximately six times greater than that of the retest scores for both the original and transformed observations. Application of the test for the equality of two correlated variances [31, pp. 195-197 and 32, pp. 190-191] indicates that the null hypothesis (17) can be rejected at the .20 level (2 tail test) for either the original or the transformed observations. Here then is an instance where the diagnostic test discussed above seems to signal the inappropriateness of assuming a test-retest correlation to be an indicator of measure reliability.

As noted earlier, the inequality of observed score variances could be due to changes in either or both of the true and error components. In turn, differences between test and retest true score variances could arise from the uniform proportional change process discussed previously or as a result of other more complex/heterogeneous types of shifts. Unfortunately, simple test-retest data are insufficient to investigate such possibilities. For the uniform proportional change case(B), it has been shown [13,pp. 453-456 and 24, p.218] that the separation of true change and measurement error requires at least three rounds of measurements. Heise [19] and Wiley and Wiley [35] have discussed procedures for estimating the reliability coefficients and proportional shift parameters when observations are available from three waves of measurements. The data examined here do not themselves suggest the occurrence of a uniform proportional change inasmuch as two of the four retest recall scores were higher than their initial test levels and two were lower. The ratios of retest to test scores ranged from .55 to 2.40 with a mean ratio of 1.17. Lacking more detailed knowledge of the test and retest conditions, we are unable to conjecture whether the instability of test and retest variances might be due to changing true score variances or non-constant error variances.

Case II

The second set of data was obtained from a packaged goods manu-
facturer and consists of pairs of recall scores for ten different
commercials featuring three brands which the firm marketed in three
separate but interrelated product classes. Here the interval between
the initial test and subsequent retest varied from two to twenty-four
months during which time the commercials had been aired with widely
differing frequencies, ranging from seven to one hundred and sixty-six
times. The first tests had also been conducted under quite disparate
circumstances. For some of the commercials the initial test had been
performed when respondents had had only one or two opportunities to be exposed to them while for others the first recall measurement was taken
after the commercials had been broadcast twenty or thirty times. Summary
statistics for these test-retest data are given in Table 2.

__INSERT_TABLE_2_HERE__

The widely ranging conditions under which these test and retest measures
were obtained represents the kind of uncontrolled situation where it would
seem unlikely that the parallel measurement model would hold. In light
of this, it is somewhat surprising to find then in Table 2 that the test-
retest corrleation has a quite respectable and statistically significant
value of .81 (p < .01, 2 tail test) and although the mean of the retest
recall scores is nearly 5 percentage points lower than the mean for the first
test, the magnitudes of the test and retest variances are very similar.
Furthermore, the results obtained from performing our suggested diagnostic
test provides no grounds for rejecting the hypothesis of equal test and
retest observed score variances (17). As indicated in Table 2, the t
statistics for testing the equality of two correlated variances is quite
small and non-significant for both the original and transformed observations.

This evidence concerning the stability of the test and retest variances is consistent with the results expected when a set of data satisfy the conditions required for a test-retest correlation to be equivalent to a reliability coefficient. To test the hypothesis of a mean difference between the test and retest scores we employ the following random effects ANOVA model used in psychometric research to estimate reliability variance components [36,pp. 124-132]:

$$(18) \quad r_{it} = \mu + C_i + V_t + \varepsilon_{it} \quad ,$$

where : r_{it} = recall score for commercial i on the tth replication of the copy test, t = 1,2 for test and retest data, respectively,

μ = grand mean,

C_i = effect of commercial i,

V_t = effect of the tth replication,

ε_{it} = error in measurement of commercial i in replication t.

The effects of commercials, replications, and errors are all assumed to be random with $C_i \sim N(0, \sigma_c^2)$, $V_t \sim N(0, \sigma_v^2)$ and $\varepsilon_{it} \sim N(0, \sigma_\varepsilon^2)$, respectively. This ANOVA model represents the case discussed in the preceding section (A) where a uniform additive change occurs in all the retest scores compared to their initial test levels. In the absence of such a shift V_t and σ_v^2 are zero and then we have the situation represented by our basic measurement model where the true commercial scores are stable across replications.

Table 3 shows the results when the above ANOVA model was applied to the arcsin transformed values of the data under consideration here.

INSERT TABLE 3 HERE

The F ratio computed from the "between tests" and "error" mean squares
in Table 3 provides a test of the hypothesis that there was a mean shift
in recall levels between the test and retest measurements.[5/] The test
yields a statistically significant F value ($p < .01$). The equality of
the test and retest variances combined with the inequality of their means
suggests the possibility that this might be an example of the Case A situation
discussed above where the test-retest correlation still represented a reliability coefficient because the difference in the test and retest means
was due to a uniform additive shift in the true scores of all the commercials.
An estimate of the shift parameter, δ in (8A), is given by the mean of the
differences between the test and retest scores for the entire set of ten
commercials -- i.e., for the arcsin transformed observations (degrees),
$\hat{\delta} = -3.75$. For the eight of the ten commercials, the retest score was
lower that the initial test value and the observed differences ($Y_{i2} - Y_{i1}$)
ranged from $+ 1.49$ to $- 10.90$ (degrees), and do not appear homogeneous.
One would, of course, be hard pressed to find a rationale for expecting
a uniform additive shift process here in the first place.

Clearly the great diversity in the conditions under which these test
and retest measures were obtained compells us to regard the notion of a
mean additive shift as only a crude method of adjusting for a heterogeneous
set of changes which cannot be estimated more precisely given only a single
pair of test-retest observations for each commercial. However, acceptance
of such an adjustment is implicit in the use of a test-retest correlation
coefficient as a measure of reliability. It can be shown that estimates
of the true and error variances which yield a reliability coefficient equal
to the test-retest correlation coefficient are those obtained by assuming
that the mean difference between test and retest scores represents a systematic source of variation and is not part of the measurement error
[36 , pp. 127-131].

This may be seen for the present example by deriving the estimates of the
variance components. Under the assumptions of the measurement model stated
at the outset, an unbiased estimate of the true score variance is obtained by
setting the between-commercials mean square equal to its expected value and
solving the resulting equation -- i.e., $64.96 = 2\sigma_\theta^2 + \sigma_\epsilon^2$. Using the estimated error variance shown in Table 3 ($\hat{\sigma}_\epsilon^2 = 7.03$), we find the estimate of the
true score variance to be 28.96, which as indicated by (14) and (14A) is equal to the

value of the covariance between the observed test and retest scores shown in Table 2. The latter is, of course, the quantity used in the numerator of (11) to compute the value of the test-retest correlation coefficient. The denominator of the correlation coefficient (11) is the geometric mean of the observed test and retest variance ($\hat{\sigma}^2_{o_1}$, $\hat{\sigma}^2_{o_2}$, respectively) which in this vase is virtually identical to the sum of the estimated true and error variances ($\hat{\sigma}_\theta + \hat{\sigma}^2_\varepsilon = 28.96 + 7.03 = 35.99$) and so the reliability index calculated from the estimated variance components ($\hat{R} = \hat{\sigma}^2_o / \hat{\sigma}^2_\theta + \hat{\sigma}^2_\varepsilon = 28.96/35.99 = .8\ 05$) has the same value as the test-retest correlation shown in Table 2.

It is important to recognize that if for some reason one were unwilling to accept the mean additive shift as a satisfactory adjustment procedure here, then the test-retest correlation coefficient could no longer be used as a reliability estimate. Suppose that one chose to disregard altogether any true score changes that might have occured between the two tests and instead assumed that all the "within commercials" mean square represented error variance as would be the case if the test and retests had been conducted under conditions that strictly conformed to the parallel measurements model. That is, assume $\hat{\sigma}^2_\varepsilon = 13.35$ rather than 7.03 as before. Proceeding in the manner described above, we would obtain a smaller estimate of the true score variance ($\hat{\sigma}^2_\theta = 25.80$) and, of course, a lower reliability ($\hat{R} = .66$).

It is interesting to note from Table 3 that the estimate of the error variance, $\hat{\sigma}^2_\varepsilon = 7.03$, obtained under the assumption of a mean shift in true scores is not too different from the theoretical value expected for the sampling variance of binomial proportion which in the arcsin scale (degrees) is equal to 821.70/n where n is the sample size. With random samples of approximately 200 respondents, if all the error variance were due to sampling a binomial process, we would expect an error variance of 821.70/200 or roughly 4. An error variance of 7.03 would compare very favorably to estimates of error variances for other on-air recall testing methods and product categories reported in Silk and Bultez [29].

One can easily imagine other situations where there would be
little or no reason to question the adequacy of the mean shift adjustment
as a means of accounting for true changes or differences between the
repeated measurements. For example, such might hold with test-retest
data arising from copy tests for a set of commercials which had been
replicated contemporaneously in a pair of cities. However, the matter
of whether such a between-replications mean difference is to be treated
as a systematic source of variation or as part of the measurement error
may be decided differently according to the type of problem or use which
the reliability information is intended to serve. For this reason, a
single summary statistic such as a test-retest correlation or a reliability
coefficient is likely to be much less useful to a copy test analyst
than the full set of results obtained from an application of the ANOVA
model (18) to the same body of data.

To summarize then, we find this set of test-retest data passed
our available diagnostic check for consistency with the assumptions
of the parallel measurement model even though the observations were
originally obtained from a wide variety of test-retest conditions
that appeared to be extremely non-equivalent for each of the commercials
individually and which were also known to differ greatly across the
entire set of commercials. At the same time, it is clear that only
by making a very strong assumption of some kind about what constitutes
measurement error here would it be possible for one to make any inference about reliability from this body of test-retest data.
Alternatively, and perhaps no more arbitrarily, one might simply
conclude that there was no defensible way to distinguish between
heterogeneous true shifts and random measurement error in these data
and therefore no inferences about reliability can be drawn from them.

DISCUSSION

Many, if not most of the test-retest correlations that have been presented in the advertising research literature as measures of the reliability of copy testing methodologies are subject to the kinds of ambiguities and limitations noted in discussing the preceding examples. To illustrate, a finding often cited is the test-retest correlation of .67 obtained from an analysis of on-air recall scores for 106 commercials first reported by Clancy and Kweskin [11] and discussed at greater length in a recent article by Clancy and Ostlund [12]. In describing the data upon which this correlation is based, the latter authors note that "The interval between test and retest for each commercial varied from one to eleven weeks" [12, p. 31]. Thus there were numerous and varied opportunities for the different test and retest samples to be effected by non-equivalent sets of influences and it is, therefore, not surprising to learn that "The average difference between the test and retest score was $\pm$ 6.4 percentage points" [12, p. 31]. The variances of the test and retest scores were not reported and so the test for their equality suggested above could not be carried out.

In order to accept Clancy and Ostlund's conclusion that the correlation of .67 indicated a "discouraging" and "modest" level of reliability, one must also be willing to assume that a uniform additive or mean difference shift was an adequate representation of whatever pattern of true differences actually developed between the times the test and retest measures were taken. The doubtfullness of the latter proposition would seem to be strongly indicated by the fact that when Clancy and Ostlund grouped the original 106 pairs of observations into eleven product categories and computed a separate test-retest correlation for each, they found that the coefficients varied widely, ranging from - .69 to .98. Similar results were obtained when they repeated the analysis with another set of test-retest data from recall studies of 32 commercials gathered by a different research organization. Clancy and Ostlund interpreted the heterogeneity of these correlations as an indication of product category variability in the reliability of the on-air testing methods. Here again, a plausible rival hypothesis is that the variability of test-retest correlation is due to the heterogeneity

of the true shifts that might have occured for the individual commercials in the
different product categories rather unstable error variances and lack of reliability
Clancy and Ostlund also seem to have been willing to entertain a
similar point of view since they report having looked into the correlation between the absolute difference in test-retest scores and
the length of the test-retest time intervals but found it to be zero.
Appel [4,5] had earlier reported finding such a relation, interpreting
the time interval as a surrogate measure of the amount of exposure
which the commercial had received between the test and retest.

It becomes important to appreciate that the lack of test-retest
correlation observed in heterogeneous collections of data like those
analyzed by Clancy and Ostlund may not simply be a reflection of random
measurement error but rather to some unascertainable extent may also be
due to true shifts induced by processes like repeated exposure and
forgetting to which a reliable and valid response measure ought to
be sensitive. By ignoring the latter possibility, one may be lead
to favor copy testing methods that generated stable results across
dissimilar conditions where differences should be found, perhaps
because the results contain a large component of non-random
measurement error such as the types of response set tendencies Appel
and Blum [3] and Wells [34] have detected in readership scores for print
ads. The folly of "stable but insensitive" copy tests was discussed
some years ago by Maloney [26] with reference to recall scores from
portfolio tests of print ads and again more recently by Weiss and
Appel [33] with reference to attitude change measures from persuasion
tests of television commercials.

An example of where a lack of correlation between repeated measures
was interpreted as being a consequence of a known source of non-equivalence
in the measures is provided by Winters and Wallace[37]. Scores from
CONPAAD tests of the same set of print ads were obtained from each of
four different groups of respondents. The size of the correlation in
the ads' scores between pairs of groups appeared to be related to the
similarity of the groups with respect to sex and age.

21

Because the test-retest correlations found in the advertising
research literature have typically been derived from post hoc analyses
of data originally gathered under dissimilar conditions for purposes
other than that of estimating measurement error, interpreting these
statistics as reliability coefficients is very suspect. Efforts to
clarify or attach meaning to such correlations are frequently frustrated
by the fact that very rarely have the reports of these statistics inc-
luded sufficient information about the test-retest data and how they
were obtained to permit the resolution of doubts. For example, on the
basis of a recent review [8] of the available published literature concerned with the reliability of various copy testing methods, it would
appear that Clancy and Ostlund [12] are unique in having provided information about the differences in test and retest means when reporting
a correlation coefficient as a reliability index.

Carrying out the aforementioned consistency checks bearing on the
stability of the distributions of the test and retest scores are small
but worthwhile steps that can be taken to remove some of the ambiguities
surrounding the interpretation of a test-retest correlation as a measure
of reliability. The needs of the practicing copy test analyst might
better be served by reliability studies if estimates of the variance
components were reported rather simply noting the test-retest correlation
coefficient. Being a ratio, the latter statistic by itself may not be
very useful for many purposes such as making reliability comparisons
between different copy testing systems or across different product
categories. Such comparisons are likely to be more meaningful if
estimates of the underlying error variances are available. In
educational and psychological research. concerned with issues of
reliability, emphasis is placed on variance components rather than
variance ratios such as a correlation or reliability coefficient [14,
24].

An alternative to estimating reliability from test-retest correlations
that has occasionally been used with reference to copy testing procedures
is to retest the same advertising stimuli not just twice, but several times
with separate samples of respondents. The advantage of this approach is

that there are more degrees of freedom available for
estimating the magnitude of error variance.[6/] Maloney
[26] has reported such data for recall scores obtained from several
repetitions of portfolio tests for print ads as have Leavitt, Waddell,
and Wells [23] and Young[38] for recall measures from on-air tests
of television commercials. Of course, here too factors other than random
measurement error may affect the stability of the several sets of retest data unless the replications are executed so as to avoid such
influences which are especially likely to arise when the multiple
retests are distributed over an extended time period and/or are conducted with samples from different populations. This is less likely
to be a problem with scores from copy testing methods that employ
"pretest-posttest" or "posttest-only control group" types of designs.
Both yield net change scores which provide some control for certain basic
threats to internal validity when repeated measurements are used to
assess reliability [10]. The coupon redemption measure used in copy
testing is derived from a posttest-control group design and results
reported from retests of the same commercials conducted over a consider-
able period of time indicated that the scores were well-behaved with
respect to measurement error [2, 21].[7/] Theater tests typically employ a pretest-
posttest design and Silk and Bultez[29] found for preference change scores
that the magnitude of the error variances estimated from retest data for several
product categories was very similar to that expected due to random sampling.

The preceding discussion has been concerned with the use of test-
retest correlations to estimate the relative magnitude of measurement
error versus true variance in copy test scores. Clearly, correlations
between repeated but non-equivalent measures may sometimes be used to
provide other kinds of assessment of the quality of measurements. For
example, Lucas computed correlations between readership scores measured
in the Advertising Research Foundation's PARM study and those produced
independently by commercial services for the same set of ads [25] .
Similarly, Appel and Blum[3] correlated readership scores obtained from
a study conducted before the ads had actually been run with those found
by a commercial service after the ads had appeared. In both these cases,

Clf
3.

the authors reported not only the correlations between the two sets
of scores but also their means, noting that the differences in means
could be partly attributed to the fact that their methodologies were
similar but not identical to those employed by the commercial services.
The correlations reported by Lucas and Appel and Blum can be interpreted as evidence bearing on the construct validity of the readership
measures [18]. In psychological research, the extent of agreement
between two separate attempts to measure the same thing is used to assess
convergent validity [9]. But measures may be invalidated for a lack
of discriminant ability as when they are found to correlate too highly
with other measures form which they were intended to differ [9]. The
correlations noted by Lucas relate to the issue of convergent validity
while that found from Appel and Blum's work bears on the question of
discriminant validity. A recent paper by Kahn and Light [22] represents one of the few attempts to assemble some empirical evidence
concerning the construct validity of copy testing methods for television
commercials.

CONCLUSION

This paper has shown how temporal shifts in the distributions of test
and retest scores affects the ability of a test-retest correlation to provide information about the relative magnitudes of "true" and "error"
variance in observed copy-testing scores and thereby serve as a concept-
ually meaningful index of reliability. It would not appear that the
limitations of test-correlations in this respect have been widely under-
stood. Application of the simple checks suggested here would make the
interpretation of these correlations less equivocal. Marketing researchers
need to recognize that a lack of test-retest correlation manifest in a
given body of data may not only be due to random measurement error but
may also be the result of "real" changes within the distributions of
scores that are purposeful and explicable. There is a danger that uncritical acceptance of test-retest correlations as measures of reliability
will lead to adoption of procedures which are "stable" but "insensitive"
to identifiable advertising variables like repetition to which a measure
should display some dependable relationship.

24

Reliability estimates free of the kinds of problems and limitations discussed here can only be expected when planned studies are undertaken for the expressed purpose of developing such information. It is beyond the scope of the present paper to offer specific proposals for such work since the requirements for it will depend upon the type of copy test decision and the type of copy test method for which reliability information is sought. The problem becomes one of experimental design in the sense of Campbell and Stanley [10]. Research is needed which employs designs which provide for control over known sources of systematic influences on copy test scores [e.g., 11, 12] which for purposes of determining random measurement error become unwanted and extraneous factors. However, the real stumbling block has not been over how reliability research should be done but whether it is worth doing. The oft-noted barriers being the high cost of the required research plus what Weiss and Appel describe as "the reluctance of management to fund research which does not address an immediate marketing problem" [32, p.59]. It is to be hoped that the present paper may stimulate a re-examination of that reluctance.

25

Table 1

SUMMARY STATISTICS FOR CASE I

	Original Observations		Arcsin Transformation	
	Initial Test	Retest	Initial Test	Retest
Mean	.240	.193	28.21	25.84
Range	.05 ∿.36	.12 ∿.25	12.92∿36.90	20.26 ∿30.00
Variance	.0197	.0032	118.76	17.81
Test-Retest Covariance	.0049		32.56	
Test-Retest Correlation Coefficient	.63 (.40 > p > .20)		.71 (.40 > p > .20)	
Test for Equality of Test and Retest Variances	t = 1.90 (.20 > p > .10)		t = 2.20 (.20 > p > .10)	
(d.f = 2, two-tail test)				

26

Table 2

SUMMARY STATISTICS FOR CASE II

	Original Observations		Arcsin Transformation	
	Initial Test	Retest	Initial Test	Retest
Mean	.193	.145	25.58	21.83
Range	.06 ∿.31	.06 ∿.27	14.18 ∿ 33.83	14.18 ∿ 31.31
Variance	.0062	.0052	37.27	34.72
Test-Retest Covariance	.0046		28.96	
Test-Retest Correlation Coefficient	.81 (p < .01)		.81 (p < .01)	
Test for Equality of Test and Retest Variances (d.f. = 8, two tail test)	t = 0.44 (n.s.)		t = 0.17 (n.s.)	

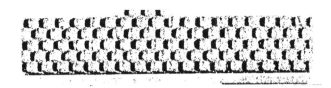

27

Table 3

ANOVA SUMMARY FOR CASE II

(Arcsin Transformation)

Source of Variation	Mean Square	d.f.	F Ratio
Between Commercials	64.96	9	$\frac{64.96}{7.03} = 9.24^{**}$
Within Commercials	13.35	10	
Between Tests Error	(70.24) (7.03)	(1) (9)	$(\frac{70.24}{7.03} = 9.99^{**})$
Total	37.80	19	

** $p < .01$

FOOTNOTES

1. The reliability index employed by Gross (12, p. 98) is the square root of R defined in (7). Such an index is also sometimes referred to in the psychometric literature [24, p. 61].

2. Morrison [27] has developed the same concept of reliability in a somewhat different but closely related manner to that set forth above. Making essentially the same assumptions as defined in (1-4) and (8-10) above, he has shown that the slope coefficient obtained by regressing the retest scores on the original test scores is equal to the ratio of the true score variance to the observed score variance -- i.e., (16) above. However, Morrison derived this relation by working with the conditional distribution of the retest scores, given the distribution of the initial tests scores. A similar analysis is given by Lord and Novick [24, pp. 64-66]. In contrast, the present approach is in terms of .the more familiar concept of a simple correlation coefficient which is the statistic reported in all of the test-retest studies of copy testing found by the present writer in the advertising research literature. Morrison's discussion also differs from the present one in that he did not emphasize empirical issues and existing studies as is done in subsequent sections of this paper.

3. If the sample sizes were known, it would, of course, be preferable to use this information and perform a weighted analysis of the observations as in [29].

4. It should be noted that Heller[20] did not report a test-retest correlation coefficient for these data. Instead he presented the original recall scores and emphasized how the test and retest studies implied different conclusions as to which commercial should be run.

5. This is equivalent to the usual "t test" for the equality of two correlated means.

29

FOOTNOTES

6. ANOVA methods for estimating reliability from multiple repeated measurements are presented in Winer [36, pp. 124-130] under the assumption that the true and error variances and hence the reliability of the measure are the same for each round of measurement. The conditions necessary for reliability to be constant across more than two sets of repeated measurements are that all observed score variances must be equal and all their covariances must also be equal. See [30] for discussion of an appropriate test of this composite hypothesis. Wiley and Wiley [35] have developed a method for estimating non-constant reliabilities from three or more waves of repeated measurements for a situation where the process of change can be modelled in terms of linear relationships between adjacent true scores and where error variances remain constant.

7. This judgement needs to be qualified because the measure used in [2, 21] is a ratio of the coupon redemption for the exposed groups to that for the control group. The sample statistics and distributions for a ratio of two random variables are complex and the variance components required to assess reliability could not be determined from the summary of results from the repeated tests presented in [2, 21].

30

REFERENCES

1. "A Synopsis of Advertising Testing Services and Procedures." Los
    Angeles: Audience Studies, Inc., no date.

2. Action Speaks Louder Than Words. Los Angeles: Tele Research, Inc.,
    1971.

3. Appel, Valentine and Milton L. Blum. "Ad Recognition and Respondent
    Set," Journal of Advertising Research, 1 (June 1961), 13-21.

4. _____. "The Reliability and Decay of Advertising Measurements,"
    Paper presented at the Marketing Conference of the National Industrial
    Conference Board, October 28, 1966.

5. _____. "On Advertising Wear Out," Journal of Advertising
    Research, 1 (February 1971), 11-13.

6. Ashton, Winifred, The Logit Transformation. London: Griffin, 1972.

7. Axelrod, Joel N. Choosing the Best Advertising Alternative. New York:
    Association of National Advertisers, Inc., 1971.

8. Bultez, Alain, Christian Derbaix, and Alvin J. Silk. "Developing and
    Testing Advertising Alternatives: Is the Magic Number One or Could it
    be Four?" Paper presented at the Fifth Annual Meeting of the European
    Academy for Advanced Research in Marketing, Fontainebleau, France,
    April, 1976.

9. Campbell, Donald T. and Donald W. Fiske. "Convergent and Discriminant
    Validation By the Multitrait-Multimethod Matrix," Psychological
    Bulletin, 56 (March 1959), 81-105.

10. _____. and Julian C. Stanley. Experimental and Quasi-
    Experimental Designs in Research. Chicago: Rand-McNally, 1963.

11. Clancy, Kevin J. and David M. Kweskin. " TV Commercial Recall Correlates,"
    Journal of Advertising Research, 11 (April 1971), 18-20.

12. _____. and Lyman E. Ostlund. "Commercial Effectiveness
    Measures," Journal of Advertising Research, 16 (February 1976), 29-36.

13. Coleman, James S. "The Mathematical Study of Change," in Hubert M.
    Blalock, Jr.and Ann B. Blalock, eds. Methodology in Social Research.
    New York: McGraw-Hill, 1971, 429-478.

14. Cronbach, Lee J., Goldine C. Gleser, Harinder Nanda, and Nageswari
    Rajaratnam. The Dependability of Behavioral Measurements. New York:
    Wiley, 1972.

15. Gross, Irwin. "Should the Advertiser Spend More on Creating Advertising?"
    Proceedings, 13th Annual Conference, Advertising Research Foundation,
    New York, 1967, 78-83.

31

16. _____. "The Creative Aspects of Advertising", Sloan Management Review, 14 (Fall 1972), 83-109.

17. Guilford, J.P. Psychometric Methods, 2nd ed. New York: McGraw-Hill, 1954.

18. Heeler Roger M. and Michael L. Ray. "Measure Validation in Marketing," Journal of Marketing Research, 9 (November 1972), 361-370.

19. Heise, David R. "Separating Reliability and Stability in Test-Retest Correlation," American Sociological Review, 34 (February 1969), 93-101.

20. Heller, Harry E. "The Ostrich and the Copy Researcher: A Comparative Analysis," Paper presented at the December 2, 1971 Meeting of the Advertising Effectiveness Research Group, New York, Chapter, American Marketing Association.

21. Jenssen, Ward J. "Sales Effects of TV, Radio and Print Advertising," Journal of Advertising Research, 6 (June 1966), 2-7.

22. Kahn, Fran and Larry Light. "Copytesting - Communication vs. Persuasion," in Advances in Consumer Research, Vol. 2, Mary Jane Schlinger, ed., Proceedings, November, 1974 Conference of the Association for Consumer Research, 595-605.

23. Leavitt, Clark, Charles Waddell, and William Wells. "Improving Day-After Recall Techniques," Journal of Advertising Research, 10 (June 1970), 13-17.

24. Lord Frederick M. and Melvin R. Novick. Statistical Theories of Mental Test Scores. Reading, Mass.: Addison-Wesley, 1968.

25. Lucas, Darrell B. "The ABC's of ARF's PARM," Journal of Marketing, 25 (July 1960), 9-20.

26. Maloney, John C. "Portfolio Tests - Are They Here to Stay?" Journal of Marketing, 25 (July 1961), 32-37.

27. Morrison, Donald G. "Reliability of Tests: A Technique Using the 'Regression to the Mean' Fallacy," Journal of Marketing Research, 10 (February 1973), 91-93.

28. Plummer, Joseph T. "Evaluating TV Commercial Tests," Journal of Advertising Research, 12 (October 1972), 21-28.

29. Silk, Alvin J. and Alain Bultez, "Product Category Variability in the Reliability of Copy Tests," Working Paper 76-48, European Institute for Advanced Studies in Management, Brussels, October 1976.

30. _____. "A Note on the Use of Tests for the Stability of Variances in Reliability Studies", Working Paper 76-53, European Institute for Advanced Studies in Management, Brussels, December 1976.

31. Snedecor, George W. and William G. Cochran, Statistical Methods, 6th ed. Ames, Iowa: Iowa State University Press, 1967.

32. Walker, Helen M. and Joseph Lev. Statistical Inference. New York : Holt, Rinehart and Winston, 1953.

. Weiss, Tibor and Valentine Appel. "Sense and Nonsense in Attitude-Change Copy Testing," _Proceedings_, 19th Annual Conference, Advertising Research Foundation, 1973, 54-59.

. Wells, William. "How Chronic Overclaimers Distort Survey Findings," _Journal of Advertising Research_, 3 (June 1963), 8-18.

. Wiley, David E. and James A. Wiley. "The Estimation of Measurement Error in Panel Data," _American Sociological Review_, 35 (February 1970), 112-117.

. Winer, B.J. _Statistical Principles in Experimental Design_. New York: McGraw-Hill, 1962.

. Winters, Lewis C. and Wallace H. Wallace. "On Operant Conditioning Techniques," _Journal of Advertising Research_, 10 (October 1970), 39-45.

. Young, Shirley. "Copy Testing Without Magic Numbers," _Journal of Advertising Research_, 12 (February 1972), 3-12.

CPSIA information can be obtained
at www.ICGtesting.com
Printed in the USA
BVHW040933290119
538940BV00016B/321/P